Module 5 Physical c and transition elemer

G000255155

Rates, equilibrium and pH

How fast?

Orders, rate equations and rate constants

Rate equations are experimentally determined and give information about the effect that the concentration of an individual reactant has on the rate of a reaction. This is expressed as the 'order' with respect to a particular reactant. A rate constant (which is temperature-dependent) can be deduced and this controls the overall rate. A reaction of order one (a first-order reaction) has a constant half-life. You must be able to explain all these terms. You must also be able to use data from the initial rate of a reaction to calculate orders of reaction and the value of the rate constant with appropriate units.

---

1 Explain what is meant by: (AO1)

a   the order of a reagent in a reaction   `1 mark`

b   the overall order of a reaction   `1 mark`

c   the rate constant   `1 mark`

d   the half-life of a first-order reaction   `1 mark`

---

2 Nitrite ions react with iodide ions in acid solution to form iodine, nitrogen monoxide and water according to the equation:

$$2NO_2^-(aq) + 2I^-(aq) + 4H^+(aq) \rightarrow 2NO(g) + I_2(s) + 2H_2O(l)$$

The rate equation for this reaction is found to be

$$rate = k[NO_2^-(aq)][I^-(aq)][H^+(aq)]^2$$

What would be the effect on the rate of the reaction if the following changes were made?

Show how you decide on your answer. (AO2)

a The concentration of $H^+(aq)$ was kept constant but the concentrations of $NO_2^-(aq)$ and $I^-(aq)$ were both doubled.

2 marks

....

....

....

b The concentration of $NO_2^-(aq)$ was kept constant but the concentrations of $H^+(aq)$ and $I^-(aq)$ were both doubled.

2 marks

....

....

....

c The concentrations of all three reagents were halved.

2 marks

....

....

....

3 Assuming the concentrations are in $mol\,dm^{-3}$ and the time is measured in seconds, give the units of the rate constant for each of the following: (AO2)

a a zero-order reaction

1 mark

....

b a first-order reaction

1 mark

....

c a reaction with an overall order of two

1 mark

....

d a reaction with an overall order of three

1 mark

....

Exam-style question

1 The following data giving the initial rate of reaction for various concentrations of NO and Br_2 are obtained for the reaction:

$$2NO(g) + Br_2(g) \rightarrow 2NOBr(g)$$

[NO]	[Br$_2$]	Initial rate/mol dm^{-3} s^{-1}
0.50	0.20	6.0×10^2
0.10	0.20	24.0
0.50	0.40	1.2×10^3

a   Determine out the orders of reaction with respect to NO(g) and Br_2(g).   **2 marks**

....

....

....

....

....

....

b   Calculate the value of the rate constant.   **2 marks**

c   Determine the units of the rate constant.   **1 mark**

....

Rate graphs and orders

You must be familiar with the shapes of the graphs obtained by plotting either concentration against time or rate against concentration as a reaction proceeds. For the former, a first-order reaction can be identified by noting that the half-life $t_{1/2}$ is constant. For a first-order reaction, the relationship $k = \ln 2/t_{1/2}$ can be used to determine the value of the rate constant. A general familiarity with the experimental methods used to obtain data about the rate of a reaction is also necessary.

4 a   Sketch concentration–time graphs for a zero-order and a first-order reaction. (AO1)

  i   zero-order reaction   ii   first-order reaction   **2 marks**

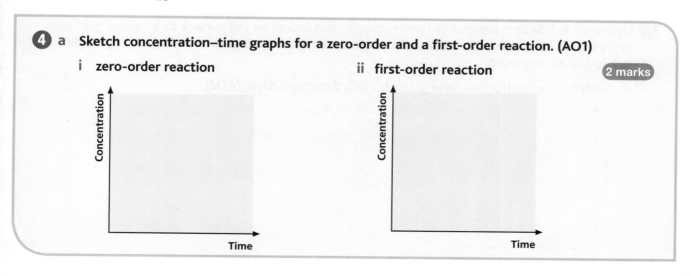

b How would you determine the initial rate of reaction from a concentration–time graph? (AO1) `1 mark`

....

c Explain what feature of the graph of a first-order reaction allows the order to be deduced? (AO1) `1 mark`

....

....

5 a Sketch rate–concentration graphs for zero-, first- and second-order reactions. (AO1) `3 marks`

   i zero-order reaction

   ii first-order reaction

   iii second-order reaction

b If the gradient of a first-order rate–concentration graph is measured, what information does this give you? Explain your answer. (AO2) `2 marks`

....

....

....

6 Hydrogen iodide decomposes to form hydrogen and iodine on the surface of a heated gold wire. Experiments have shown that the reaction proceeds at a constant rate until all the hydrogen iodide has decomposed.

   a Sketch a concentration–time graph for this decomposition. (AO2) `1 mark`

b   Give the rate equation for the decomposition. (AO1)   `1 mark`

c   What are the units of the rate constant for the decomposition? (AO2)   `1 mark`

7 A substance with an original concentration of 2.00 mol dm⁻³ takes 120 s to decompose to a concentration of 1.00 mol dm⁻³.

a   i   If the decomposition is first-order, what is the value of the rate constant? Show your working below. (AO2)   `3 marks`

    ii How long would it take for an initial concentration of 0.800 mol dm⁻³ to reduce to 0.100 mol dm⁻³? (AO2)   `1 mark`

b   i   If the decomposition is zero-order, what is the value of the rate constant? Show your working below. (AO2)   `2 marks`

    ii In this case, how long would it take for an initial concentration of 0.800 mol dm⁻³ to reduce to 0.100 mol dm⁻³? (AO2)   `2 marks`

8 In an experiment to determine the initial rate of a reaction, the time t taken to reach a certain point in the reaction is sometimes measured. The rate is then considered to be proportional to $1/t$.

Explain what assumption is being made in this rate measurement. (AO3)   `1 mark`

How valid is this assumption likely to be? (AO3)   `1 mark`

9 The progress of a reaction can sometimes be followed by measuring a change in a physical property that occurs during the course of a reaction.

   a Why is the measurement of a physical property usually preferable to a chemical analysis of the reaction mixture during the reaction? (AO3)   **2 marks**

....

....

....

   b Suggest a *physical* property that might be used to follow the progress of each of the following reactions. Explain your answers. (AO3)

     i The reaction of gaseous hydrogen and bromine to form gaseous hydrogen bromide.   **2 marks**

....

....

     ii The reaction of magnesium carbonate with dilute hydrochloric acid to form carbon dioxide and a solution of magnesium chloride.   **2 marks**

....

....

---

⑦

Exam-style question

1 The following table gives some data for the breakdown of sulfuryl chloride (SO_2Cl_2) into sulfur dioxide and chlorine over two hours.

Time/min	0	20	40	60	80	100	120
$[SO_2Cl_2]$/mol dm^{-3}	1.50	1.13	0.848	0.636	0.477	0.358	0.268

   a Plot a concentration–time graph for the decomposition of sulfuryl chloride. **3 marks**

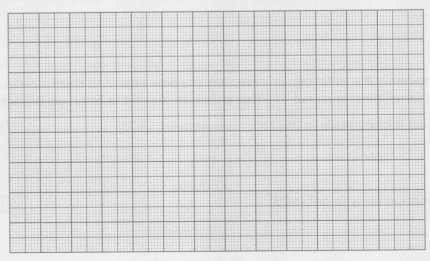

   b Use your graph to show that the decomposition of sulfuryl chloride is first-order.

Show any working that you use below.

`2 marks`

c Calculate the value of the rate constant, k, and give its units.

`2 marks`

Rate-determining step

Reactions often proceed by a number of steps before the products are obtained. Usually only one step is controlling the overall rate of the reaction, and this is called the rate-determining step. You must know how the orders of reaction relate to the rate-determining step and how this provides a clue as to the mechanism of the reaction that is taking place.

10 a Hydrogen and iodine(I) chloride react to form hydrogen chloride and iodine. Write an equation for this reaction (AO2)

`1 mark`

b In an experiment, it is found that the orders with respect to both hydrogen and iodine(I) chloride are 1. Give the rate equation for this reaction. (AO1)

`1 mark`

c What information can you deduce about the rate-determining step? (AO1)

`1 mark`

11 Nitrogen(v) oxide decomposes to form nitrogen(IV) oxide and oxygen. The rate equation for this decomposition is:

rate = $k[N_2O_5]$

a Write an equation for the decomposition. (AO2)

`1 mark`

b Suggest a two-step mechanism for the decomposition that is consistent with the rate equation. (AO3)

`2 marks`

Exam-style question

1 The following data were obtained from an initial rate experiment between nitrogen(II) oxide and hydrogen to form nitrogen and steam.

Experiment number	[NO]/mol dm^{-3}	[H$_2$]/mol dm^{-3}	Initial rate of reaction /mol dm^{-3} s^{-1}
1	0.250	0.100	9.50×10^{-3}
2	0.375	0.100	2.14×10^{-2}
3	0.500	0.200	7.60×10^{-2}

a Use the data to determine the orders of reaction for nitrogen(II) oxide and hydrogen.

4 marks

b i Write the rate equation for the reaction.

1 mark

ii Calculate a value for the rate constant k, and give its units.

3 marks

units of k

c i Write an overall equation for the reaction

1 mark

ii Suggest a two-step mechanism for the decomposition that is consistent with the rate equation.

2 marks

Effect of temperature on rate constants

The value of the rate constant is dependent on temperature, and the relationship is given by the Arrhenius equation, $k = Ae^{-E_a/RT}$. This equation will be provided on the exam data sheet, and questions will focus on its use.

⑫ State the meaning of the following terms in the Arrhenius equation. (AO1)   **4 marks**

k ....

E_a ....

T ....

R ....

⑬ a   If a set of results is obtained for the value of the rate constant at a set of different temperatures, *explain* what graph should be plotted in order to obtain values for E_a and A. (AO2)   **2 marks**

....

....

....

....

....

b   Explain how the graph would be used to obtain a value for E_a.   **1 mark**

....

....

....

⑭ What is the value of the activation energy in kJ for a reaction at 600°C that has a pre-exponential factor of 3.00×10^5 and a rate constant of $6.50 \times 10^{-3} \, dm^3 \, mol^{-1} \, s^{-1}$? Give your answer to three significant figures. (AO2)   **4 marks**

15 At 400 K, the rate constant for a reaction is $7.2 \times 10^{-5}\,s^{-1}$. At 600 K, the rate constant for the same reaction is $3.4 \times 10^{-3}\,s^{-1}$.

a   Determine the activation energy of the reaction. Give your answer in kJ to two significant figures. (AO2)

3 marks

b   Calculate the value of the pre-exponential factor, A. (AO2)

2 marks

15

Exam-style question

1 At high temperatures, cyclopropane can be converted into propene.

Cyclopropane   Propene

C_3H_6 C_3H_6

Measurements are made of the value of the rate constant at various temperatures. These results are shown in the table below.

Experiment	1	2	3	4	5
Temperature/°C	450	500	550	600	650
Temperature, T/K					
$(1/T)$/K^{-1}					
Rate constant, k/s^{-1}	9.35×10^{-6}	1.90×10^{-4}	2.68×10^{-3}	2.79×10^{-2}	2.25×10^{-1}
$\ln k$					

a   Complete the table to give values of the temperature T in K and also both 1/T and $\ln k$ to three significant figures.

3 marks

b   Plot a graph 1/*T* (*x*-axis) against ln *k* (*y*-axis).   `2 marks`

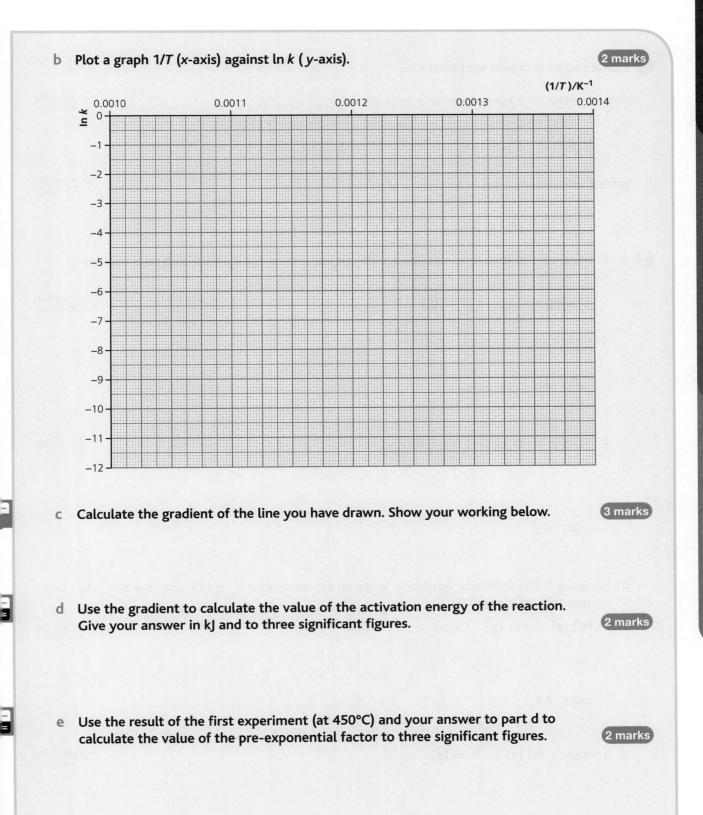

c   Calculate the gradient of the line you have drawn. Show your working below.   `3 marks`

d   Use the gradient to calculate the value of the activation energy of the reaction.
Give your answer in kJ and to three significant figures.   `2 marks`

e   Use the result of the first experiment (at 450°C) and your answer to part d to
calculate the value of the pre-exponential factor to three significant figures.   `2 marks`

How far?

Equilibrium

This section extends the ideas of equilibrium met in Module 3. You must be able to write equilibrium constants in terms of concentrations and also in terms of partial pressures. Calculations of quantities present at equilibrium, given appropriate data, are required and the qualitative effect of a change in conditions must be understood.

1 Explain what is meant by: (AO1)

mole fraction of a component in a mixture

1 mark

....

....

partial pressure of a gas in a mixture of gases

1 mark

....

....

2 a For each of the following equilibria, write an expression for K_c. If the units are $mol\,dm^{-3}$, what will be the units of K_c? (AO1)

   i $2NO(g) + O_2(g) \rightleftharpoons 2NO_2(g)$

2 marks

units of K_c

....

   ii $5CO(g) + I_2O_5(s) \rightleftharpoons I_2(g) + 5CO_2(g)$

2 marks

units of K_c

....

b For each of the following equilibria, write an expression for K_p. If the units are kPa, what will be the units of K_p? (AO1)

   i $PCl_5(g) \rightleftharpoons PCl_3(g) + Cl_2(g)$

2 marks

units of K_p

....

   ii $N_2(g) + 3H_2(g) \rightleftharpoons 2NH_3(g)$

2 marks

units of K_p

....

3 In an experiment, 0.500 mol $PCl_5(g)$ and 0.500 mol $PCl_3(g)$ were mixed in a flask of volume 600 cm³ and allowed to reach the following equilibrium:

$$PCl_5(g) \rightleftharpoons PCl_3(g) + Cl_2(g)$$

The equilibrium mixture was found to contain 0.400 mol $PCl_5(g)$.

a   Calculate K_c. (AO2)

b   Explain what happens to the value of K_c and to the position of the equilibrium if the pressure in the flask is increased. (AO3)

....

....

....

....

....

....

4 If in the equilibrium mixture

$N_2(g) + 3H_2(g) \rightleftharpoons 2NH_3(g)$

the mole fraction of $N_2(g)$ is 0.42 and the mole fraction of ammonia is 0.26, what is the mole fraction of hydrogen? (AO1)

....

....

5 At a high temperature, for the equilibrium

$2SO_3(g) \rightleftharpoons 2SO_2(g) + O_2(g)$

K_p is 8.2 kPa.

In an equilibrium mixture, the partial pressure of SO_2 is measured as 40 kPa and the partial pressure of O_2 is 10 kPa. What is the partial pressure of SO_3? (AO2)

6 When 0.40 mol of $P_4(g)$ is heated, the following equilibrium is formed:

$$P_4(g) \rightleftharpoons 2P_2(g)$$

It is found that the equilibrium mixture contains 0.35 mol of P_4 and the pressure in the container is 40 kPa. Calculate the value of K_p. (AO2)

7 marks

20

Exam-style questions

1 a A student heats 0.50 mol of HI(g) in a 2.0 dm³ flask to a temperature of 730 K. An equilibrium is formed:

10

$$2HI(g) \rightleftharpoons H_2(g) + I_2(g)$$

An analysis of the equilibrium mixture shows that it contains 0.055 mol of iodine.

i Suggest in outline how the amount of iodine in the equilibrium mixture might be determined.

2 marks

ii Calculate the value of K_c. Give your answer to an appropriate number of significant figures.

4 marks

b When K_c is calculated at 900 K, its value is found to be greater than it is at 730 K. Explain whether the decomposition of HI(g) is exothermic or endothermic.   2 marks

....

....

....

....

2 At 3000 K, for the equilibrium

$$2H_2O(g) \rightleftharpoons 2H_2(g) + O_2(g)$$

K_p is 9.50×10^{-4} kPa.

In an experiment, 18.0 g of water is heated to 3000 K and allowed to come to equilibrium. Analysis shows that the equilibrium mixture obtained contains 10.8 g of steam.

a Write an expression for K_p.   1 mark

b **i** Calculate the amounts in moles of each component of the mixture once the equilibrium has been formed.   2 marks

ii Calculate the partial pressures of each component in the equilibrium mixture in terms of the total pressure P_T.   3 marks

Acids, bases and buffers

Brønsted–Lowry acids and bases

The Brønsted–Lowry theory defines acids and bases in the context of a reaction and in terms of conjugate acid–base pairs. Calculations of pH are required for strong and weak monobasic acids and strong alkalis. The limitations of the approximations made in these calculations should be understood. An understanding of buffer solutions and the calculation of their pH are required. An ability to sketch the pH changes occurring during titrations and an appreciation of the appropriate choice of indicator to select is also expected.

1 a State what is meant by a Brønsted–Lowry acid. (AO1) `1 mark`

b i Explain what is meant by a conjugate acid–base pair. Illustrate your answer with an example. (AO1) `2 marks`

ii When concentrated sulfuric acid is added to concentrated nitric acid, an acid–base reaction takes place and the solution obtained contains an HSO_4^- ion.

Write an equation for the reaction that takes place and label the conjugate acid–base pairs that are formed, showing which are the acids and which are the bases. (AO3) `2 marks`

2 a Write equations for the following reactions. Include state symbols. (AO1) `3 marks`

i dilute nitric acid and aqueous calcium hydroxide

ii dilute sulfuric acid and aluminium oxide solid

iii ethanoic acid and magnesium carbonate solid

b Write ionic equations for each of the above reactions in part a. Include state symbols. (AO1)   **3 marks**

i

....

ii

....

iii

....

3 Give an expression for K_a for the ionisation of methanoic acid. (AO1)   **1 mark**

If the concentrations are measured in $mol\,dm^{-3}$, what are the units of K_a? (AO1)   **1 mark**

....

What is meant by pK_a? (AO1)   **1 mark**

....

pH and [H⁺(aq)]

4 a i What is meant by pH? (AO1)   **1 mark**

....

ii Calculate the pH of a monobasic acid for which $[H^+] = 3.45 \times 10^{-3}\,mol\,dm^{-3}$. (AO1)   **1 mark**

b Calculate the $[H^+]$ for a monobasic acid whose pH is 4.70 (AO2)   **1 mark**

5 a Calculate the pH of a $0.050\,mol\,dm^{-3}$ solution of hydrochloric acid. (AO2)   **1 mark**

b Calculate the pH of $0.050\,mol\,dm^{-3}$ aqueous sodium hydroxide. (AO2)   **2 marks**

c   i   Calculate the pH of a $5.0 \times 10^{-3}\,mol\,dm^{-3}$ solution of methanoic acid.
        For methanoic acid, K_a is $1.6 \times 10^{-4}\,mol\,dm^{-3}$. (AO2)
        2 marks

    ii State *two* assumptions that have been made in calculating the pH of the
        methanoic acid. (AO3)
        2 marks

        ....

        ....

d   Explain why the pH of a $1.0 \times 10^{-8}\,mol\,dm^{-3}$ aqueous solution of hydrochloric
    acid is *not* equal to 8? (AO3)
    2 marks

    ....

    ....

    ....

Buffers: action, uses and calculations

6 a   Explain what is meant by a buffer solution. (AO1)
    1 mark

    ....

    ....

b   Explain why the pH of a solution of ethanoic acid rises when sodium ethanoate
    is added to it. (AO2)
    2 marks

    ....

    ....

    ....

    ....

c   Explain how the conjugate acid–base pair of ethanoic acid and ethanoate ions acts
    as a buffer solution when a few drops of acid are added to it. (AO2)
    2 marks

    ....

    ....

    ....

    ....

d Explain how the conjugate acid–base pair of ethanoic acid and ethanoate ions acts as a buffer solution when a few drops of alkali are added to it. (AO2) **2 marks**

....

....

....

....

7 a Calculate the pH of a buffer solution made by adding 25.0 g of sodium methanoate to 1.0 dm³ of 0.500 mol dm⁻³ solution of methanoic acid. For methanoic acid, K_a is 1.6×10^{-4} mol dm⁻³. (AO2) **4 marks**

b Explain *three* approximations that you have made in carrying out the calculation in part a. (AO3) **3 marks**

....

....

....

....

....

8 Explain how blood pH is controlled by a carbonic acid–hydrogencarbonate buffer system. (AO2) **2 marks**

....

....

....

....

....

Neutralisation

9 On the grid below, sketch the shapes of the graphs of pH (*y*-axis) against volume of alkali added (*x*-axis) that would be obtained for the following titrations. (AO1)

a   A strong acid of pH = 1.0 against a strong base of pH 11.0. Assume that the end-point occurs when 20.0 cm³ of the strong base has been added.   **2 marks**

b   A weak acid of pH = 3.0 against a strong base of pH = 11.0. Assume that the weak acid is twice as concentrated as the strong acid.   **2 marks**

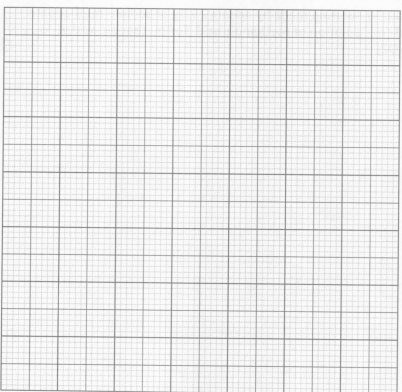

10 a   In aqueous solution, a weak acid HA forms an equilibrium with its ions, H^+(aq) and A^-(aq). The colour of HA(aq) is blue, while the ion A^-(aq) is yellow.

Explain how HA could be used as an acid–base indicator. (AO3)   **4 marks**

....

....

....

....

....

....

....

....

....

....

....

b If the equilibrium constant for HA is $1.8 \times 10^{-4}\,mol\,dm^{-3}$, calculate the pH of the indicator when in aqueous solution it has equal numbers of HA(aq) molecules and A^-(aq) ions. (AO2)

2 marks

c Suggest whether HA would be a suitable indicator for a titration between a weak alkali and a strong acid. Explain your answer. (AO3)

1 mark

....

....

....

11 Describe the procedure that is required to use a pH meter to measure the pH of a solution. (AO2)

4 marks

....

....

....

....

....

....

....

Exam-style questions

(20)

1 In an experiment, a mixture is made by adding $12.5\,cm^3$ of $2.00\,mol\,dm^{-3}$ aqueous sodium hydroxide to $25.0\,cm^3$ of $2.00\,mol\,dm^{-3}$ ethanoic acid.

(10)

a i State the *three* ions or molecules in highest concentration in the mixture that is obtained.

2 marks

....

....

....

ii Calculate the concentration of ethanoic acid in the mixture. Show your working.

2 marks

iii Calculate the concentration of ethanoate ions in the mixture. Explain your answer.

1 mark

....

....

....

b How does this mixture act as a buffer solution when further drops of sodium hydroxide are added?

3 marks

....

....

....

....

c Calculate the pH of the buffer solution. For ethanoic acid $K_a = 1.7 \times 10^{-5}\,mol\,dm^{-3}$.

2 marks

2 a i Calculate the pH of a $0.020\,mol\,dm^{-3}$ solution of propanoic acid. For propanoic acid $K_a = 1.3 \times 10^{-5}\,mol\,dm^{-3}$.

10

2 marks

ii Calculate the pH of $0.020\,mol\,dm^{-3}$ sodium hydroxide.

2 marks

b On a separate sheet of graph paper, sketch the titration curve of pH against volume of alkali that would be obtained by adding 30.0 cm³ of 0.020 mol dm⁻³ sodium hydroxide to 20.0 cm³ of 0.020 mol dm⁻³ propanoic acid. (AO2)

3 marks

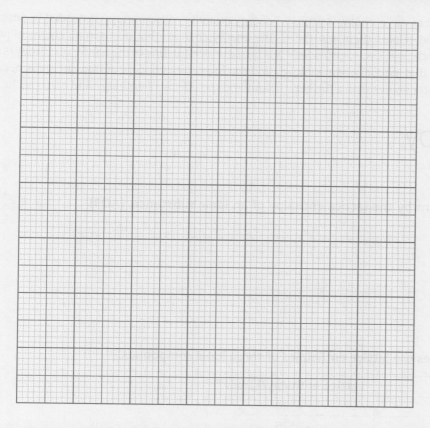

c Which, if any, of the indicators in the following table would be suitable use in this titration? Explain your answer.

2 marks

Indicator	pH range of the indicator
Methyl red	4.2–6.3
Bromocresol purple	5.2–6.8
Thymol blue	8.0–9.6
Alizarin yellow	10.1–13.0

....

....

....

....

....

d If aqueous ammonia had been used for the titration instead of sodium hydroxide, which, if any, of the indicators in the table above would be suitable for use in the titration? Explain your answer.

2 marks

....

....

....

....

Energy

This section introduces a number of enthalpy terms, particularly lattice enthalpies, and relates them via the Born–Haber cycle. Another enthalpy cycle provides information about the changes that occur when ionic solids are added to water. The latter shows that a further energy quantity called entropy is needed to describe fully the energy changes that occur in all chemical processes. A broad understanding of the nature of entropy is required, and the important concept of free energy is necessary to combine the effects of enthalpy and entropy changes.

Lattice enthalpy

1 **a** **Define what is meant by the lattice enthalpy of an ionic substance. (AO1)** `3 marks`

....

....

....

....

b **The table below gives the lattice enthalpies of three ionic compounds.**

Compound	Lattice enthalpy/kJ mol^{-1}
Sodium chloride	−780
Magnesium chloride	−2526
Magnesium oxide	−3791

i **Explain why the lattice enthalpies of sodium chloride and magnesium chloride differ in their value. (AO2, AO3)** `3 marks`

....

....

....

....

....

ii **Explain why the lattice enthalpies of magnesium chloride and magnesium oxide differ in their value. (AO2, AO3)** `3 marks`

....

....

....

....

....

....

....

Born–Haber and related enthalpy cycles

2 Define the following terms. (AO1)

enthalpy change of atomisation

....

....

second ionisation enthalpy

....

....

first electron affinity

2 marks

....

....

3 Use the following information to construct a Born–Haber cycle and use it to calculate the lattice enthalpy of magnesium chloride. (AO2)

4 marks

Enthalpy term	Enthalpy change/kJ mol^{-1}
Enthalpy change of atomisation of magnesium	+149
Enthalpy change of atomisation of chlorine	+122
First ionisation enthalpy of magnesium	+740
Second ionisation enthalpy of magnesium	+1500
Electron affinity of chlorine	−364
Enthalpy change of formation of magnesium chloride	−642

....

....

4 Define the following terms. (AO1)

enthalpy change of hydration

....

....

enthalpy change of solution

....

....

5 Use the following information to construct an enthalpy cycle and use it to determine the enthalpy of solution of zinc chloride. (AO2)

Enthalpy term	Enthalpy change/kJ mol^{-1}
Lattice enthalpy of zinc chloride	−2734
Enthalpy change of hydration of Zn^{2+}	−2105
Enthalpy change of hydration of Cl^-	−338

$$Zn^{2+}(g) + 2Cl^-(g)$$

Increasing enthalpy

Exam-style question

1 a i Define standard enthalpy of formation. `2 marks`

....

....

ii Define lattice enthalpy. `2 marks`

....

....

b The table below gives enthalpy changes of formation of sodium chloride and potassium chloride.

Substance	Enthalpy change of formation/kJ mol^{-1}
Sodium chloride	−411
Potassium chloride	−437

The following table gives data for the enthalpy change of atomisation and the first ionisation enthalpies of sodium and potassium.

Substance	Enthalpy change of atomisation/kJ mol^{-1}	First ionisation enthalpy/kJ mol^{-1}
Sodium	+107	+496
Potassium	+89.2	+419

Use these data to calculate by how much the values of the lattice enthalpies of sodium chloride and potassium chloride differ. `4 marks`

c Explain why the lattice enthalpies of sodium chloride and potassium chloride differ. `2 marks`

....

....

....

....

d   If the lattice enthalpy of sodium chloride is −781 kJ mol⁻¹, construct an enthalpy
    cycle and determine the enthalpy of solution of sodium chloride.   **3 marks**

$$\Delta_{hyd}H(Na) = -418 \text{ kJ mol}^{-1}; \Delta_{hyd}H(Cl) = -338 \text{ kJ mol}^{-1}$$

Enthalpy and entropy

Entropy

1 a   Explain qualitatively what you understand by the term 'entropy'. (AO1)   **1 mark**

b   Explain whether ice or water will have the greater entropy. (AO3)   **2 marks**

c   For each of the following reactions, *explain* whether the products or the reactants will have
    the higher total entropy. (AO2)

    i $Zn(s) + H_2SO_4(aq) \rightarrow ZnSO_4(aq) + H_2(g)$

2 marks

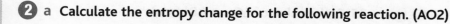

    ii $C(s) + CO_2(g) \rightarrow 2CO(g)$

2 marks

    iii $2NO(g) + O_2(g) \rightarrow 2NO_2(g)$

2 marks

2   **a**   Calculate the entropy change for the following reaction. (AO2)

2 marks

     $C_2H_4(g) + 3O_2(g) \rightarrow 2CO_2(g) + 2H_2O(l)$

     $S^{\ominus}(C_2H_4(g)) = 219.5\,J\,mol^{-1}\,K^{-1}$; $S^{\ominus}(O_2(g)) = 204.9\,J\,mol^{-1}\,K^{-1}$;
     $S^{\ominus}(CO_2(g)) = 213.8\,J\,mol^{-1}\,K^{-1}$; $S^{\ominus}(H_2O(l)) = 70.0\,J\,mol^{-1}\,K^{-1}$

  b   Explain whether the sign of the answer to part a is predictable. (AO3)

2 marks

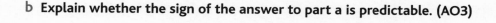

Free energy

3   **a**   What is meant by describing a reaction as 'feasible'? (AO1)

1 mark

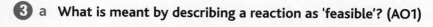

  b   Why do reactions that are feasible not always occur readily? (AO2)

2 marks

c State the relationship between ΔG, ΔH and ΔS. (AO1) 1 mark

d For each of the following reactions, *explain* whether raising the temperature of the reaction will make it more feasible. (AO2)

   i a reaction for which both ΔH and ΔS are positive   2 marks

   ii a reaction for which ΔH is positive and ΔS is negative   2 marks

 4 For the reaction $N_2O_4(g) \rightarrow 2NO_2(g)$, at 25°C, the value of $\Delta G = +4.8\,kJ\,mol^{-1}$.

At 100°C, $\Delta G = -8.4\,kJ\,mol^{-1}$.

Calculate the value of ΔS for the reaction. (AO2)   4 marks

Exam-style question

 1 a i Calculate the entropy change that occurs when 1 mol of graphite is oxidised to carbon dioxide.   1 mark

$S^{\ominus}$(graphite) = 5.7 J mol^{-1} K^{-1}; $S^{\ominus}$(O$_2$(g)) = 204.9 J mol^{-1} K^{-1}; $S^{\ominus}$(CO$_2$(g)) = 213.8 J mol^{-1} K^{-1}

ii Explain whether this reaction is feasible at room temperature and pressure. `3 marks`

....

....

....

....

b i Calculate $\Delta H^\ominus$ for the following reaction, under standard conditions. `1 mark`

$$4CuO(s) \rightarrow 2Cu_2O(s) + O_2(g)$$

$$\Delta H^\ominus(CuO(s)) = -155.2\,kJ\,mol^{-1};\ \Delta H^\ominus(Cu_2O(s)) = -166.7\,kJ\,mol^{-1}$$

ii $\Delta S^\ominus$ for the same reaction is $+232.5\,J\,mol^{-1}\,K^{-1}$.

Calculate the temperature in °C at which the reaction becomes just feasible. `4 marks`

iii What assumption have you made in carrying out the calculation in part b ii? `1 mark`

....

....

Redox and electrode potentials

An explanation of the terms oxidising agent and reducing agent and the construction of ionic equations for redox reactions is required. Redox reactions can form the basis of titrations, and calculations from results of titrations involving MnO_4^- ions, $I_2/S_2O_3^{2-}$ and other redox reactions are expected.

The methods for the determination of standard electrode potentials and the use of these potentials to calculate cell potentials and to predict the feasibility of reactions must be understood. Application of the principles of electrode potentials to modern storage cells follows.

Redox

1 Explain what is meant by an oxidising agent in terms of electron transfer. (AO1) `1 mark`

....

....

2 Use half ionic equations to construct overall equations for the following reactions, and state which is the reducing agent in each reaction. (AO2)

a $Cl_2(aq) + Fe^{2+}(aq) \rightarrow Cl^-(aq) + Fe^{3+}(aq)$

4 marks

....

....

....

Reducing agent is:

....

b $Cu(s) + NO_3^-(aq) + H^+(aq) \rightarrow Cu^{2+}(aq) + NO(g) + H_2O(l)$

4 marks

....

....

....

Reducing agent is:

....

c $VO_2^+(aq) + H^+(aq) + I^-(aq) \rightarrow VO^{2+}(aq) + I_2(aq) + H_2O(l)$

4 marks

....

....

....

Reducing agent is:

....

d $Cr_2O_7^{2-}(aq) + H^+(aq) + SO_2(aq) \rightarrow Cr^{3+}(aq) + SO_4^{2-}(aq) + H_2O(l)$

4 marks

....

....

....

Reducing agent is:

....

Redox titrations

3 a Write an equation for the reaction of $Fe^{2+}(aq)$ with acidified $MnO_4^-(aq)$. (AO2)

2 marks

....

....

b Calculate the concentration in $mol\,dm^{-3}$ of a solution containing $Fe^{2+}(aq)$ if 25.0 cm³ of the solution requires 23.7 cm³ of acidified aqueous 0.0200 mol dm⁻³ $KMnO_4$ to reach the end-point in a titration. (AO2)

2 marks

c The Fe^{2+}(aq) was made by dissolving y grams of iron(II) sulfate crystals ($FeSO_4 \cdot 7H_2O$) to make 250 cm³ of solution. Calculate the value of y. (AO2)   **2 marks**

d Describe in detail how the solution in part c could be made in the laboratory. (AO3)   **4 marks**

....

....

....

....

....

....

....

....

e What is the reason for acidifying the $KMnO_4$ solution for the titration? (AO3)   **1 mark**

....

....

....

4 A solution contains an ion, XO_3^-(aq), with a concentration of 0.100 mol dm⁻³.

It is found that 25.0 cm³ of this solution when acidified reacts exactly with 37.5 cm³ of a solution containing 0.200 mol dm⁻³ of Fe^{2+}(aq) ions.

a Calculate the oxidation number of X when the reaction is complete. (AO2)   **3 marks**

b Write an equation for the reaction that has taken place. (AO2)   **2 marks**

....

....

....

5 Solution A contains dichromate ions, $Cr_2O_7^{2-}(aq)$, at a concentration of x mol dm^{-3}.

Solution B contains bismuthate ions, $BiO_3^-(aq)$, also at a concentration of x mol dm^{-3}.

Both solutions are titrated separately against 25.0 cm^3 of a solution containing $Fe^{2+}(aq)$ to which excess sulfuric acid has been added.

It is found that 8.0 cm^3 of solution A are required to oxidise the $Fe^{2+}(aq)$ to $Fe^{3+}(aq)$. In the reaction, all the $Cr_2O_7^{2-}(aq)$ ions are reduced to $Cr^{3+}(aq)$.

It is also found that 24.0 cm^3 of solution B are required to oxidise the $Fe^{2+}(aq)$ to $Fe^{3+}(aq)$. In the reaction, all the $BiO_3^-(aq)$ ions are reduced to an ion Bi^{n+}.

a Write an equation for the reaction of $Fe^{2+}(aq)$ and acidified $Cr_2O_7^{2-}(aq)$. (AO2)　`2 marks`

b Deduce the formula of the ion formed when $BiO_3^-(aq)$ is reduced in its reaction with $Fe^{2+}(aq)$. (AO3)　`2 marks`

c Write an equation for the reaction of $Fe^{2+}(aq)$ and acidified $BiO_3^-(aq)$. (AO2)　`2 marks`

　　　　　　　　　　　　　　　　　　　　　　　　　　　　　　⏱ **10**

Exam-style question

1 In an experiment, 3.638 g of compound GIO_3 (where G is a metal ion) was dissolved in distilled water to make 250.0 cm^3 of solution. Excess iodide was added to 25.0 cm^3 of the solution of $GIO_3(aq)$ and iodine was formed.

The iodine was then titrated against 20.0 cm^3 of 0.600 mol dm^{-3} aqueous sodium thiosulfate.

It was found that 20.0 cm^3 of the thiosulfate was required to reach the end-point.

a Write an equation to show how $IO_3^-(aq)$ reacts with $I^-(aq)$ to form iodine.　`2 marks`

b   Write the equation for the reaction of iodine and thiosulfate ions. **1 mark**

....

c   Use the equations in parts a and b to determine how many moles of thiosulfate ions would be required for every 1 mol of $IO_3^-(aq)$ that reacted with iodide ions. **1 mark**

....
....
....

d   Use your answer to part c to determine the amount in moles of $IO_3^-(aq)$ in 250 cm³ of the solution made by dissolving the 3.638 g of compound GIO_3. **3 marks**

e   Use your answer to part d to calculate the relative atomic mass of G. Hence identify the metal, G. **2 marks**

....

Electrode potentials

6 a   Describe the $Cl_2(g)/2Cl^-(aq)$ half-cell. (AO1) **2 marks**

....
....
....

b Draw a fully labelled diagram illustrating the determination of the standard electrode potential for the $Fe^{3+}(aq)/Fe^{2+}(aq)$ half-cell. (AO1)

5 marks

7 Calculate the standard cell potential obtained from the combination of the following half-cells. (AO2)

a $Mn^{2+}(aq)/Mn(s)$ and $Ni^{2+}(aq)/Ni(s)$

1 mark

$Mn^{2+}(aq) + 2e^- \rightleftharpoons Mn(s); E^\ominus = -1.19\,V$

$Ni^{2+}(aq) + 2e^- \rightleftharpoons Ni(s); E^\ominus = -0.25\,V$

b $Cr^{3+}(aq)/Cr(s)$ and $Ag^+(aq)/Ag(s)$

1 mark

$Cr^{3+}(aq) + 3e^- \rightleftharpoons Cr(s); E^\ominus = -0.74\,V$

$Ag^+(aq) + e^- \rightleftharpoons Ag(s); E^\ominus = +0.80\,V$

c $Br_2(aq)/2Br^-(aq)$ and $MnO_4^-, H^+(aq)/Mn^{2+}(aq)$

1 mark

$Br_2(aq) + 2e^- \rightleftharpoons 2Br^-(aq); E^\ominus = +1.09\,V$

$MnO_4^-(aq) + 8H^+(aq) + 5e^- \rightleftharpoons Mn^{2+}(aq) + 4H_2O(l); E^\ominus = +1.51\,V$

8 For each of the following, use standard electrode potentials to decide whether a reaction with $Fe^{3+}(aq)$ will be feasible under standard conditions. Explain how you made your decision, and write an equation for the reaction (if any) that occurs. (AO2)

$$Fe^{3+}(aq) + e^- \rightleftharpoons Fe^{2+}(aq); \; E^\ominus = +0.77\,V$$

a $Fe^{3+}(aq)$ and $Sn^{2+}(aq)$ **2 marks**

$$Sn^{4+}(aq) + 2e^- \rightleftharpoons Sn^{2+}(aq); \; E^\ominus = +0.15\,V$$

b $Fe^{3+}(aq)$ and $Br^-(aq)$ **2 marks**

$$Br_2(aq) + 2e^- \rightleftharpoons 2Br^-(aq); \; E^\ominus = +1.09\,V$$

9 a Explain why the reaction between $Fe^{2+}(aq)$ and $Ag^+(aq)$ is feasible under standard conditions, and give an equation for the reaction. (AO2) **2 marks**

$$Fe^{3+}(aq) + e^- \rightleftharpoons Fe^{2+}(aq); \; E^\ominus = +0.77\,V$$
$$Ag^+(aq) + e^- \rightleftharpoons Ag(s); \; E^\ominus = +0.80\,V$$

b Describe what you would observe when $Fe^{2+}(aq)$ is added to $Ag^+(aq)$. (AO1) **2 marks**

c Is it possible to decide from these data whether this reaction will proceed slowly? Explain your answer. (AO3) **2 marks**

Storage and fuel cells

10 Electrochemical cells for use in mobile phones are based on the transport of lithium ions. These have many advantages, but can involve a risk if the phone is damaged.

Name *two* risks that might occur if the battery is damaged. (AO1) **2 marks**

Exam-style question

1 a i Draw a fully labelled diagram to illustrate the measurement of the standard electrode potential of $Ni^{2+}(aq)/Ni(s)$. **5 marks**

ii Apart from completing the circuit, explain fully the function of the salt bridge in the circuit. **3 marks**

....

....

....

iii The standard electrode potential for $Ni^{2+}(aq)/Ni(s)$ is $-0.25\,V$. In which direction will the electrons flow through the external circuit you have drawn? Explain your answer. **2 marks**

....

....

b i The nickel/zinc storage cell operates in alkaline conditions. The standard electrode potential for the nickel electrode of the cell is

$$2NiOOH + 2H_2O + 2e^- \rightleftharpoons 2Ni(OH)_2 + 2OH^-; \quad E^\ominus = +0.49\,V$$

The reaction that is occurring at the zinc electrode during discharge of the cell involves the precipitation of zinc hydroxide as a result of the oxidation of zinc metal in the presence of hydroxide ions. No other products are formed.

Write the half-equation for this process. **1 mark**

....

ii The maximum voltage obtainable from this cell when operated under standard conditions is $1.73\,V$.

Calculate the standard electrode potential for the reaction at the zinc electrode. Explain how you obtained your answer. **2 marks**

....

....

iii The cell is rechargeable. Write an overall equation for the reaction that is occurring during the recharging of the cell. **1 mark**

....

Transition elements

Transition elements

This section requires a knowledge of the electron configuration of transition metal atoms and an appreciation of their similarities in terms of redox chemistry, formation of coloured ions and catalytic behaviour. In addition, an understanding of how elements and ions form complex ions using ligands and the resultant stereochemistry of four- and six-coordinated complexes is included.

Transition metal ions can react via ligand exchange reactions and by forming characteristically coloured hydroxide precipitates. These reactions are useful in identifying a solution containing an unknown ion. The redox reactions of transition elements are extensive and you need to be familiar with some of the interconversions between ions.

Properties

1 State the numbers of protons, electrons and neutrons in each of the following, and write the electron configuration for each. (AO1).   `3 marks`

$^{56}_{26}Fe$

....

$^{55}_{25}Mn^{2+}$

....

$^{65}_{29}Cu^+$

....

2 Explain the difference between the terms 'd-block element' and 'transition element'. (AO1)   `2 marks`

....

....

....

....

3 State *four* typical properties of transition metals other than their electrical conductivity and their high melting and boiling points. In each case give an example that illustrates the properties you describe. (AO1)   `4 marks`

....

....

....

....

....

....

....

Ligands and complex ions

4 Explain what is meant by each of the following. (AO1)

a ligand

2 marks

....

....

a complex ion

2 marks

....

....

coordination number

2 marks

....

....

5 Give an example of an octahedral complex ion, and draw its structure. (AO2)

2 marks

....

6 Give the charge of the metal ion in the following complex ions. (AO1)

4 marks

$[Fe(CN)_6]^{3-}$

$[Cu(NH_2CH_2CH_2NH_2)_3]^{2+}$

$[Ag(S_2O_3)_2]^{3-}$

$[Co(NH_3)_4(Cl)_2]^+$

7 Complexes with four-fold coordination can either be tetrahedral or planar in shape.

a Name and draw an example of an ion with a tetrahedral shape. (AO1, AO2)

2 marks

....

b i Ni²⁺ ions form a planar complex ion with four cyanide ions (CN⁻). Draw the planar complex formed. (AO2)

`1 mark`

ii Platin exhibits stereoisomerism. Explain this type of stereoisomerism, and illustrate your answer by drawing the structures of platin. (AO2)

`2 marks`

....
....
....
....
....

iii Explain the role of *cis*-platin as an anti-cancer drug. (AO1)

`2 marks`

....
....
....
....

43

Exam-style question

1 **a** Give the name and formula of a bidentate ligand, and explain how it forms a complex ion. (AO2)

3 marks

....

....

....

....

b Complexes involving bidentate ligands exhibit optical isomerism.

i What structural property does a molecule have to possess that explains why optical isomers form?

1 mark

....

....

ii Draw a diagram to illustrate optical isomerism for the ion $[Ni(H_2NCH_2CH_2NH_2)_3]^{2+}$.

2 marks

Ligand substitution

8 Many reactions involve ligand substitutions. Give *two* examples of ligand substitution reactions involving copper ions. In each case, describe what you would observe as the reaction took place, and give an equation for the reaction. (AO1, AO2)

5 marks

a Example 1

....

....

....

....

b **Example 2**

....

....

....

9 a Outline the importance of iron in haemoglobin in the transportation of oxygen in the human body. (AO1) `4 marks`

....

....

....

....

....

....

b Explain how the inhalation of carbon monoxide affects the transportation of oxygen in the human body. (AO1) `2 marks`

....

....

....

....

Precipitation reactions

10 a An aqueous solution of cobalt chloride ($CoCl_2$) is pink in colour and undergoes the reactions shown below. In each case, state what type of reaction has taken place and suggest a formula for the product obtained. (AO3)

i When aqueous sodium hydroxide is added, a pale blue solid is formed in the solution. `1 mark`

....

....

ii When concentrated hydrochloric acid is added, a deep blue solution is formed and the reaction is equivalent to the reaction that occurs with aqueous copper chloride. `1 mark`

....

....

b i When aqueous ammonia is added to aqueous cobalt chloride, at first a pale blue precipitate is formed, but when an excess of aqueous ammonia is added, the precipitate dissolves to form a solution containing the complex ion $[Co(NH_3)_6]^{2+}$.

Construct ionic equations, including state symbols, for the reaction of aqueous cobalt chloride and ammonia. (AO2) `2 marks`

....

....

45

ii If hydrogen peroxide is added to $[Co(NH_3)_6]^{2+}$, it rapidly converts into $[Co(NH_3)_6]^{3+}$ and hydroxide ions are also produced.

Construct an ionic equation for the reaction of aqueous $[Co(NH_3)_6]^{2+}$ and hydrogen peroxide. (AO3)

1 mark

....

....

Redox reactions

11 Excess aqueous potassium iodide is added to 25.0 cm³ of a solution of iron(III) sulfate. Iodine is produced and this is titrated against 0.100 mol dm⁻³ sodium thiosulfate.

It is found that 23.7 cm³ of the sodium thiosulfate is required to react with all of the iodine produced.

a Write equations for the reactions that are taking place. (AO2)

2 marks

....

....

b Calculate the concentration of the iron(III) sulfate solution in mol dm⁻³. (AO2, AO3)

3 marks

12 When copper(I) sulfate is added to dilute sulfuric acid, a disproportionation reaction takes place.

a What do you understand by the term 'disproportionation'? (AO1)

1 mark

....

....

b Describe what you would observe, and write an ionic equation for the reaction. (AO1)

2 marks

....

....

....

Qualitative analysis

Tests for ions

1 Describe how you could distinguish between the following solutions in the laboratory. Give the results of any tests that you use. (AO2)

   a $FeCl_2(aq)$ and $FeBr_2(aq)$ `2 marks`

....

....

   b $FeCl_2(aq)$ and $MnCl_2(aq)$ `2 marks`

....

....

....

Exam-style question

`15`

1 **a** Describe and explain how aqueous iron(II) chloride and aqueous chromium(III) chloride could be distinguished by their chemical reactions when aqueous sodium hydroxide is slowly added until present in excess.

     Provide ionic equations for the reactions you use.   `4 marks`

....

....

....

....

....

....

....

   b **i** Outline how you could obtain a solution containing $Cr_2O_7^{2-}(aq)$ from a solution of $Cr^{3+}(aq)$ ions and give ionic equations for the reactions that occur.   `5 marks`

....

....

....

....

....

....

....

In an experiment, a solution containing 1.585 g of chromium(III) chloride is reacted to form a solution containing $Cr_2O_7^{2-}(aq)$. Once the solution has been purified so that it only contains $Cr_2O_7^{2-}(aq)$, the volume of the solution produced is measured to be 85.0 cm³.

Then 25.0 cm³ of the solution containing $Cr_2O_7^{2-}(aq)$ ions is titrated against a 0.100 mol dm⁻³ solution of $Fe^{2+}(aq)$. It is found that 21.6 cm³ of the $Fe^{2+}(aq)$ is required to react exactly with the $Cr_2O_7^{2-}(aq)$ ions. In this titration, the $Cr_2O_7^{2-}(aq)$ ions are reduced back to $Cr^{3+}(aq)$ ions.

ii Construct an equation for the reaction of $Cr_2O_7^{2-}(aq)$ ions and $Fe^{2+}(aq)$. `2 marks`

....

....

....

iii Calculate the percentage conversion of $Cr^{3+}(aq)$ to $Cr_2O_7^{2-}(aq)$ that has taken place. `5 marks`

Philip Allan, an imprint of Hodder Education, an Hachette UK company, Blenheim Court, George Street, Banbury, Oxfordshire OX16 5BH

Orders
Bookpoint Ltd, 130 Milton Park, Abingdon, Oxfordshire OX14 4SB
tel: 01235 827827
fax: 01235 400401
e-mail: education@bookpoint.co.uk
Lines are open 9.00 a.m.–5.00 p.m., Monday to Saturday, with a 24-hour message answering service. You can also order through the Philip Allan website: www.philipallan.co.uk
© John Older and Mike Smith 2016
ISBN 978-1-4718-4735-6
First printed 2016

Impression number   5   4   3   2
Year   2020   2019   2018   2017

Cover photo reproduced by permission of Fotolia
Typeset by Aptara, Inc.

Printed in Dubai

Hachette UK's policy is to use papers that are natural, renewable and recyclable products and made from wood grown in sustainable forests. The logging and manufacturing processes are expected to conform to the environmental regulations of the country of origin.